Textiles in Filtration

R.Senthil Kumar

A.P (SRG),

Department of Textile Technology,

Kumaraguru College of Technology,

Coimbatore-49, India

sen29iit@yahoo.co.in

Filtration Textiles - Contents

1.1 Introduction

1.2 Filtration – Market scenario

1.3 Filtration – Definition & Terms

1.4 Filtration – Principles of particle retention

1.5 Filtration – Collection Efficiency

1.6 Theory of Filtration

    1.6.1 Filtration – Fundamentals

    1.6.2 Filtration Resistance

    1.6.3 Fluid Flow through porous media

    1.6.4 Particle settlement

1.7 Filtration – Classification

1.8 Filter Media

    1.8.1 Factors Considered For Filter medium Selection

    1.8.2 Fibres used in dry and wet filtration

    1.8.3 Filter fabric – Yarn and Fabric Structure

    1.8.4 Property requirement of Filter fabric

    1.8.5 Criteria to evaluate filter media

    1.8.6 Operational aspects of filter fabric media

    1.8.7 Filter fabric selection procedure

1.9 Nonwoven Filter fabrics

    1.9.1 Dry laid filter fabrics

    1.9.2 Wet laid filter fabrics

    1.9.3 Characteristics of Nonwoven filter media

    1.9.4 Factors influencing properties of Nonwoven filter

1.10 Filter ratings

    1.10.1 Absolute rating

    1.10.2 Nominal rating

    1.10.3 Mean filter rating

    1.10.4 Beta ratio

1.10.5 Filter efficiency

1.11 Applications

Filtration Textiles

Introduction:

There is hardly a human activity, industrial, commercial or domestic, that is not affected by filtration. The general importance of environmental protection justifies mention of the topic here – because filtration has a major role to play in many of the schemes trying to achieve this protection. The market forces exerted by the imposition of environmental legislation are an important driver for the filtration market. Textile filter fabrics are an essential part of countless industrial processes, contributing to product purity, savings in energy/production costs and a cleaner environment.

Filtration – Market Share:

The filtration industry today is a diverse and technically sophisticated business with annual sales reported to be in excess of $100 billion. The filtration business has evolved over time to become a complex industry with very specific requirements for each area of use. Performance standards for the media used in virtually every application have become very stringent. The usage of filtration and similar separation equipment is quite evenly spread throughout the economy, with the two largest end-use sectors being those in which the largest number of individual filters is found. The domestic and commercial sector with its many water filters (and coffee filters, and suction cleaner filters) is one; and the transport system sector with its huge number of engine filters for intake air, fuels and coolants, is the other.

Filtration – Definition:

Basically, filtration means the capture and retention of small particles from a moving stream of either gas or liquid, with minimum resistant to flow. Filtration conditions vary widely and also the equipment for filtration and type of filter used.

Filtration – Terms and Definition:

a) Pressure Drop (Δp): Pressure drop through a filter is defined by the following expression:

$$\Delta p = P1 - P2$$

Where P1 is the pressure before the filtration and P2 is the pressure after the filtration.

b) Filter Efficiency E: The filter efficiency is defined as a ratio between the quantity of particles retained in the filter and the number of dispersed particles found in the suspension.

c) Filter Capacity Q: Filter capacity is defined by the amount of particles deposited in it [expressed in g or kg] and that accumulated before a drop in pressure begins. The capacity of a filter must be specified for each particles size.

d) Cleaning Efficiency: It is the ratio of dust retained by fabric after cleaning to total dust deposited expressed in percentage.

e) Degree of Filtration: This parameter defines the ratio between a certain size particles that enters the filter and the particles of the same size that leave the filter.

f) Porosity: It is the ratio of the volume of voids to the volume of fabric.

$$\text{Porosity} = \frac{(\text{Volume of fabric} - \text{Volume of fiber})}{\text{Volume of fabric}} \times 100$$

$$\text{Porosity} = 1 - \frac{\text{Fabric density}}{\text{Fibre density}} \times 100$$

Filtration – Principles of particle retention:

Filtration of particles relies on any one or more of the following principles:

   a) Impaction

   b) Diffusion

   c) Straining

d) Electrostatics

e) Sedimentation

f) Interception

Particles can be influenced by any one of these principles, or all of them simultaneously.

a) Impaction:

As large particles move along with an air stream, their inertias prevent them from making abrupt changes in direction. If an obstruction such as a series of water droplets or fibers (glass, foam rubber, cloth, etc.) is placed randomly across the air stream path, there is a certain probability that a given particle will collide with the obstruction. As particle size and the number of particles increase, so does the probability of collision.

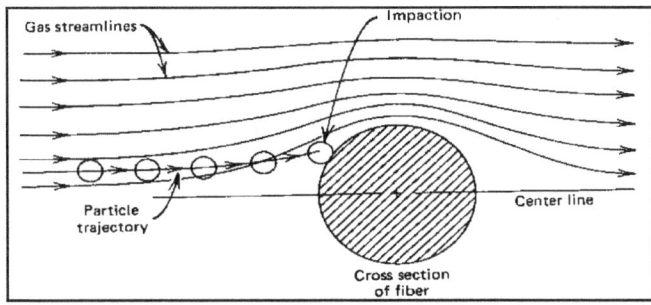

Fig 5.1 – Particle Retention by Impaction

Thus the efficiency of removing particles from an air stream by impaction is a function of particle size, fiber size and the number of fibers. The greater the number of fibers in the filter fabric (thus the deeper the bed, the higher the pressure drop), the higher the filtration efficiency. In turn, as the dust particles collect, they themselves become part of the filter media, thereby increasing efficiency by adding to the number of possible collisions for other suspended particles. As the collected particles build up on the filter, system pressure drop increases, usually a good indication that the path through the filter, the equivalent filter depth, is increasing. This principle of filtration is most commonly found in fiber filters and certain wet collectors.

b) Diffusion:

When particles become very small, their mass is so low that, should they collide with any air molecules, just the random motion of the air molecules will cause them to rebound randomly. This motion is commonly referred to as Brownian movement.

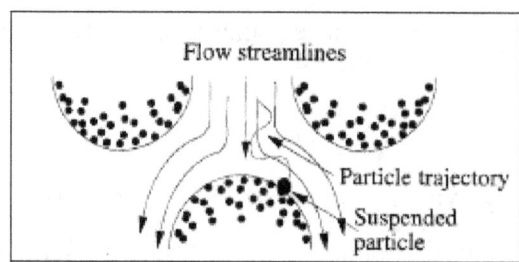

Fig 5.2 – Particle Retention by Diffusion

If the velocity of the air stream is low, this diffusion movement will in turn cause random collisions with fiber or droplets in the way of the airflow. Hence, much like impaction, probabilities can be developed for collisions due to this diffusion. Key factors are fiber size, fiber quantity and air stream velocity. As with impaction, as more and more particles collect, the probability of collision (efficiency) for other particles is enhanced, but with an associated increase in pressure drop.

c) Straining

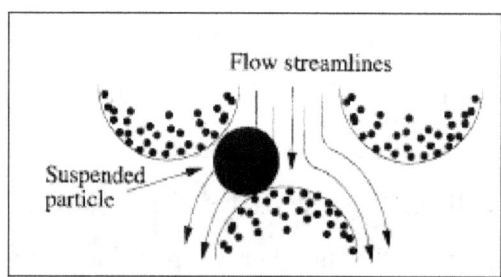

Fig 5.3 –Particle Retention by Straining

If the width of a passage is smaller than that of the particle suspended in the air stream, then the particle will be stopped and held. However, as each particle plugs a hole, air resistance increases. Standard house screens are typical of this filter type. Small particles pass through it but bugs cannot pass through. Very small particles are seldom collected using this method which is primarily used only for specialized laboratory experiments.

d) Electrostatics

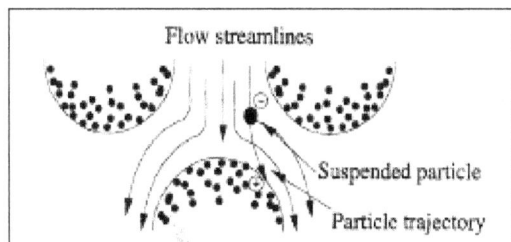

Fig.5.4 – Particle Retention by Electrostatic attraction

If a charged particle passes through an electrostatic field it is attracted to an oppositely charged body. Such charges can be generated and imparted to particles in an air stream in much the same way as static charges develop during the combing of one's hair or just walking across a rug. Electrons are stripped from large quantities of molecules with the net effect that particles of dirt not otherwise collected might be charged by friction as they pass through, then collected as they attach themselves to oppositely charged bodies. This effect can occur inside such filtering devices as fiber beds which operate primarily on the principles of impaction and diffusion but have their efficiencies enhanced by electrostatic effects.

Charges may be purposely induced onto air stream particles by applying energy to a special configuration of wires and plates stretched across the air stream. These devices, called electrostatic precipitators, form a special category of air filtration. Whether particle charges are induced by applying energy to a dirty air stream or occur naturally, they can be valuable tools in increasing air cleaning effectiveness.

e) Sedimentation:

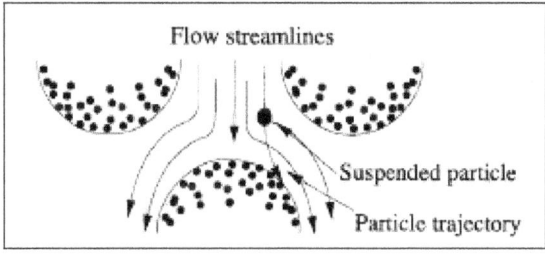

Fig.5.5-Particle Retention by Sedimentation

When the fluid flow is directed downwards through a filter, gravitational sedimentation effects will cause particles to settle vertically through the flow streamlines, as the latter distort around the collector.

f) Interception:

If the suspended particle radius is greater than the distance between the flow streamline which contains the particle and the collecting media grain, then the suspended particle will contact the

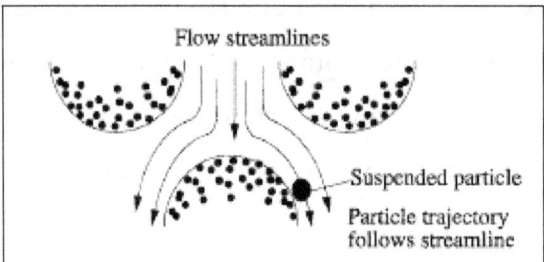

Fig.5.6 – Particle Retention by Interception

target, in the absence of any repulsive mechanisms.

Filtration - Collection Efficiency:

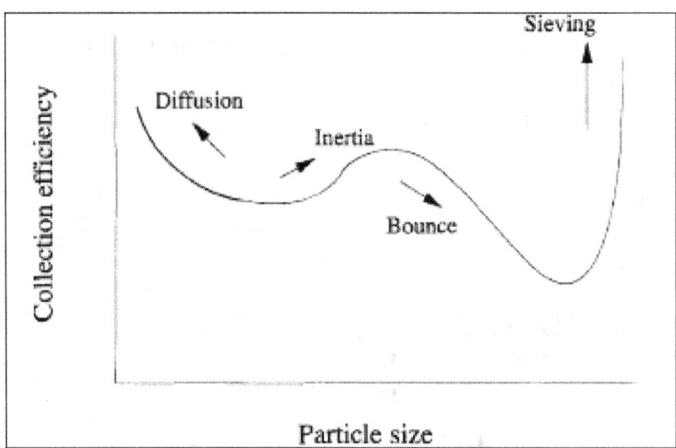

Fig.5.7 - Capture efficiency as a function of particle size in a deep bed filter

The graph relating deep-bed filtration efficiency and the size of suspended particles can be explained in terms of the relative importance of diffusion, inertia and straining. At low particle diameter removal efficiency is mainly due to diffusion. This effect becomes less relevant at higher diameters. However, as particle size increases inertial impaction becomes more relevant and efficiency increases again with size. Eventually straining, or sieving, becomes the dominant mechanism.

Filtration Fundamentals:

A successful selection procedure is closely linked to the proper choice of the medium to be used in the separation. A large proportion of industrial-scale process difficulties relate to the interaction between the impinging particles and the pores in the filter medium. The ideal circumstance, where all separated particles are retained on the surface of a medium is often not realized; particle penetration into cloth or membrane pores leads to an increase in the resistance of the medium to the flow of filtrate. This process can ensue to the level of total blockage of the system such difficulties can be avoided, if the pores in the medium are all smaller than the smallest particulate in the mixture processed.

The filtrate velocity V, through the clean filter medium is proportional to the pressure differential ΔP imposed over the medium; the velocity is inversely proportional to the viscosity of the flowing liquid μ and the resistance of the medium. These relationships may be expressed mathematically as:

$$\mathbf{V_0 = \Delta P / \mu R_m}$$

Filter cake resistances vary over a wide range, from free filtering sand-like particulates to high resistance sewage sludges. Generally, the smaller the particle, the higher will be the cake resistance.

Fluid Flow through Porous Media

The fundamental relation between the pressure drop and the flow rate of liquid passing through a packed bed of solids, such as that shown in below figure, was first reported by Darcy in 1856. The liquid passes through the open space between the particles, i.e. the pores or voids within the bed. As it flows over the surface of the solid packing frictional losses lead to a pressure drop. The amount of solids inside the bed is clearly important; the greater this is the larger will be the surface over which liquid flows and, therefore, the higher the pressure drop will be as a result of friction. The volume available for fluid flow is called the porosity or voidage, and this is defined below.

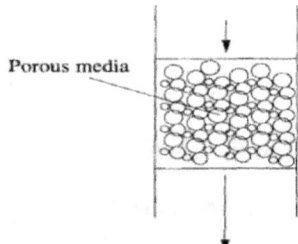

Fig 5.8 - Schematic diagram of porous media

In many solid-liquid separations the use of solid concentration is often preferred to porosity. This is usually the volume fraction of solids present within the bed C; porosity is the void volume fraction so these two fractions sum to unity. Hence solid volume fraction concentration is:

$$C = 1 - \varepsilon$$

Darcy discovered that the pressure loss was directly proportional to the flow rate of the fluid.

Particle settlement:

The critical factor in the creation and maintenance of dust-laden air or particle contaminated water is the nature of the particle, in terms of its size and its relative density. This is because of the development of a constant, terminal velocity by a particle falling freely through a fluid, whose value, as demonstrated by Stokes law, is:

- directly proportional to the square of the particle diameter
- directly proportional to the difference between particle and fluid densities
- indirectly proportional to fluid viscosity.

This relationship is strictly true only for spherical particles falling while isolated from one another by some distance. Non-spherical particles are accounted for by the use of an effective diameter (often determined by the reverse process of measuring a terminal velocity and calculating back to the diameter). Stokes law holds well enough in the case of fluids with a low solids concentration such as concerns a study of contaminant removal.

If the particle size is small enough so that the terminal velocity is very low, or if there are random movements in the fluid at velocities in excess of this terminal figure, then the particle effectively does not settle out, and a stable suspension results – which may then need decontaminating. The settling velocity in air of a spherical particle, whose specific gravity is 1, is given by: $1.8 \times 10^{-5} \times d^2$ metres per minute where d is the particle diameter, expressed in µm.

Filtration –Types:

a) Wet Filtration:

Filter fabrics used for separation of solid particles from liquids in the form of cake are known as wet filtration. In wet filtration free flow of liquid through media is not restricted but solid particles are easily stopped through the textile used.

b) Dry Filtration

In Dry Filtration, the dusts are removed by using bag filters. Large numbers of nonwoven or woven bags are used and by the means of fan pure air is filtered out.

Classification of filtration

Depending on the process of separation, filtration is classified as:

a) Particle filtration

Particle filtration is the separation of particles having size above 10 microns. These can be filtered out easily without any usage of micro porous membrane.

b) Microfiltration

Microfiltration is a filtration process, which removes contaminants from a fluid (liquid & gas) by passage through a micro porous membrane. A typical microfiltration membrane pore size range is 0.1 to 10 microns (µm). Microfiltration is fundamentally different from reverse osmosis and nano-filtration because those systems use pressure as a means of forcing water to go from low pressure to high pressure. Microfiltration can use a pressurized system but it does not need to include pressure.

c) Ultra filtration

Ultra filtration (UF) is a variety of membrane filtration in which hydrostatic pressure forces a liquid against a semi permeable membrane. Suspended solids and solutes of high molecular weight are retained, while water and low molecular weight solutes pass through the membrane.

This separation process is used in industry and research for purifying and concentrating macromolecular solutions, especially protein solutions. Ultra filtration is not fundamentally different from microfiltration or nano-filtration, except in terms of the size of the molecules it retains. Mostly, ultra filtration is applied in cross-flow mode and separation in ultra filtration undergoes concentration polarisation.

d) Nano-filtration

Nano-filtration is a relatively recent membrane filtration process used most often with low total dissolved solids water such as surface water and fresh groundwater, with the purpose of softening (polyvalent cation removal) and removal of disinfection by-product precursors such as natural organic matter and synthetic organic matter. Nano-filtration is also becoming more widely used in food processing applications such as dairy, for simultaneous concentration and partial (monovalent ion) demineralisation. Nano-filtration (NF) is a cross-flow filtration technology which ranges somewhere between ultra filtration (UF) and reverse osmosis (RO). The nominal pore size of the membrane is typically below 1 nanometer, thus Nano-filtration. Nano-filter membranes are typically rated by molecular weight cut-off (MWCO) rather than nominal pore size. The trans-membrane pressure (pressure drop across the membrane) is required considerably lower than the one used for RO, reducing the operating cost significantly. However, NF membranes are still subject to scaling and fouling and often modifiers such as anti-scalants are required for use.

e) Reverse osmosis

Reverse osmosis is similar to the membrane filtration treatment process. However, there are key differences between reverse osmosis and filtration. The predominant removal mechanism in membrane filtration is straining, or size exclusion, so the process can theoretically achieve perfect exclusion of particles regardless of operational parameters such as influent pressure and concentration. RO (Reverse Osmosis) however involves a diffusive mechanism so that separation efficiency is dependent on influent solute concentration, pressure and water flux rate. It works by using pressure to force a solution through a membrane, retaining the solute on one side and allowing the pure solvent to pass to the other side. This is the reverse of the normal osmosis process, which is the natural movement of solvent from an area of low solute concentration, through a membrane, to an area of high solute concentration when no external pressure is applied. Nonwoven wetlaid polyester substrates support reverse osmosis membranes in spiral wrap modules in a $30 million worldwide nonwovens market. The modules are found in systems predominantly located in arid regions where seawater is converted to potable water. Spunbond fabrics are used as pleat supports and separators in virtually every microporous membrane cartridge sold, accounting for nonwoven sales of approximately $35 million per year.

Methods of filtration

There are many different methods of filtration; all aim to attain the separation of substances. Separation is achieved by some form of interaction between the substance or objects to be removed and the filter. The substance that is to pass through the filter must be a fluid, ie, a liquid or gas. Methods vary depending on the location of the targeted material, ie, whether it is in the fluid phase or not.

a) Solid Gas separation

Example: Filters used in cigars, filters used in AC systems.

b) Solid Liquid separation

Example: Filters used in sewage disposal plants, filters in chemical industries, water purifiers, etc.

Filter Media:

A filter is any device in which a separation is achieved among other components of a suspension or solution, in a fluid - which may be a liquid or a gas - where the separation is caused by mechanical means, without the involvement of a change in phase (such as the melting of a solid, or the evaporation.. Filtration is almost entirely a characteristic of the size of the particle, droplet or molecule being separated. The successful performance of a filter is largely dependent on the selection of a suitable filter medium. The filtration mechanisms invoked in separations using such media will depend mainly on the mode of separation. Thus in "cake" filtration, ideally, impinging particles should be larger than the pores in the medium. The filtration efficiency of various filter medium are given below:

Table 5.1 - Types of filter media and particle retention characteristic		
Main type	**Sub-divisions**	**Smallest particle retained (μm)**
Solid fabrications	(a) flat wedge-wire screens (b) wire-bound tubes (c) stacks of rings	100 10 5
Rigid porous media	(a) ceramics and stoneware (b) carbon (c) plastics (d) sintered metals	1 1 1 <1
Cartridges	(a) sheet fabrications (b) yarn wound (c) bonded beds	3 5 1
Metal sheets	(a) perforated (b) woven wire	100 5
Plastic sheets	(a) woven monofilaments (b) fibrillated film (c) porous sheets	10 5 0.1
Woven fabrics		5
Link fabrics		200
Non-woven media	(a) filter sheets (b) felts and media felts (c) paper media - cellulose	0.5 10 5

	- glass (d) bonded media	2 0.1
Loose media	(a) fibres (b) powder	< 1 < 0.1

A successful filter medium is likely to be required to combine many different properties, ranging from its filtration characteristics and its chemical resistance to its mechanical strength, the dimensions in which it is available, and its wettability. The following factors relating to the filtration medium must be considered in the selection of textile filter:

- ❖ Composition
- ❖ Temperature
- ❖ Nature of Particulate Matter
- ❖ Fineness of Particulate Matter
- ❖ Concentration of Particulate Matter
- ❖ Loading on the Filter Surface

The fabric should process greatest possible collection efficiency, low pressure collection efficiency, low pressure drop, small filtering area, low penetration of dust in the fabric and cost. In the case of fabric filters, the fabric performs very little of actual filtering, it provides substratum or matrix for the primary dust cake to form, which in turns collects the particulate and allows air to flow through the fabric. So the fabric should be able to permit the development of a loose and porous cake on its surface and also to release the cake during cleaning. As far as fabric is concerned, its abrasion resistance, chemical resistance, tensile strength and permeability should be considered.

Filter Media Design/Selection criteria:

The primary factors which influence the design or selection of filter media may be listed as thermal and chemical conditions, filtration requirements, equipment consideration and cost. The selection of the type and fineness of the fibre is largely governed by the following circumstances prevailing in filtration:

- Temperature

- Humidity
- Chemical conditions
- Composition and size distribution of dust particles

a) Thermal and Chemical Conditions

The thermal and chemical conditions of the liquid being filtered effectively determine the type of polymer which is used in fibre/filament production. Historically, filter fabrics were produced by weaving yarns spun from natural fibres such as cotton which, on wetting, would swell to produce highly efficient media. Cotton fabrics are used extensively where gas temperature is below 80°C and acid gases are absent. However, on the debit side, in chemically aggressive conditions, their life expectancy is somewhat limited. Wool fabrics are more resistant to acid than cotton and are used in the collection of the metallurgical fumes and for fine abrasive dust such as cement.

Fibres and their properties						
Fibres	Density g/cc	Max. Operating Temp °C	Resistance to:			
			Acids	Alkali	Oxidising Agents	Hydrolysis
Polypropylene	0.91	95	VG	VG	P	G
Polyethylene	0.95	85	VG	VG	P	G
Polyester (PBT)	1.28	100	G	P	F	P
Polyester (PET)	1.38	100	G	P	F	P
Polyamide 6.6	1.14	110	P	G	P	F
Polyamide 11	1.04	100	P	G	P	F
Polyamide 12	1.02	100	P	G	P	F
PVDC	1.70	85	VG	G	VG	G
PVDF	1.78	100	VG	VG	G	VG
PTFE	2.10	150+	VG	VG	VG	VG
PPS	1.37	150+	VG	VG	F	VG
PVC	1.37	80	VG	VG	F	VG
PEEK	1.30	150+	G	G	F	VG
VG = Very good		**G = Good**		**F = Fair**		**P = Poor**

By comparison, synthetic fibres are generally much more durable but even so, it is still important to make the correct selection for the conditions which prevail in the filter. The excellent mechanical, physical and chemical properties of synthetic fibres offer high performance characteristics in the filtration process itself.

The use of filter cloth made of synthetic fibres brings the following advantages:

- Greater filtrate purity and improved hygiene condition of filtration process.
- Reduced fabric weight owing to the higher strength of constituent materials.
- More efficient rinsing of the filter cloth in the filtering systems and washing in the washing machines.
- Easier and more rapid drying.
- Full resistance to rot during the out of operation of the filtering system.
- Better resistant to effects of elevated temperature and moisture.

If the operation temperature does not exceed 150°C polyester is generally used. The most widely used fibre in dry filter media is polyester (approx.70%). If the resistance to hydrolysis of polyester is inadequate then acrylic is used. Polyamide fibres, for example, will not tolerate continuous exposure to strong acids and, conversely, polyester fibres will degrade when exposed to strong bases and prolonged hydrolytic conditions. Polyamides yarns exhibit good abrasion resistance but are sensitive to acid particulate. Aliphatic polyamides are of only major importance as polyester performs equally well at lower cost. Aromatic polyamides (aramid) have gained some importance in the production of nonwoven filter materials with their high heat resistance.

Polypropylene is relatively inert to both acids and bases, and hence is the most widely used polymer in liquid filtration, it too has an "Achilles heel" in its susceptibility to attack from oxidizing agents. The presence of chlorine or heavy metal salt is the potential source of such attack.Polypropylene with its low thermal resistance is little used. PTFE (polytetrafluoroethyelene) is virtually resistant to chemicals, has a maximum service temperature of 280°C. PTFE fibres of course are resistant to most agents but carry a cost premium which in most cases is prohibitive.

b) Filtration Requirements

In order to satisfactorily fulfil filtration requirements, the ideal filter medium will provide

i) Resistance to chemical/mechanical attrition:

Polymer selection in relation to chemical conditions was discussed already. Similarly mechanical conditions, such as the abrasive nature of the slurry and the tensile forces acting on the fabric will be items to consider when selecting the appropriate yarns and the density of the thread spacing in the fabric. The abrasive forces in this context arise from the shape and nature of the particles in the slurry. Materials with sharp edges tend to abrade internally, fibre/filament breakage and ultimately lead to pinhole formation in the filter fabric. The designing of filter fabric should be done in such a way to withstand the impact of such forces.

ii) Resistance to blinding:

Closing of the filter medium pores resulting in reduced gas flow or an increased pressure drop across the medium is termed as "Blinding".This is a well-known term which relates to particulate matter becoming trapped, sometimes irretrievably, within the interstices of the fabric and ultimately leading to a serious reduction in throughput. Fabric blinding may be temporary or permanent. If cloth may be rejuvenated by washing, either externally or in-situ, then the blinding is temporary. If medium cannot be cleaned easily and the pores opened, it is called as "permanent blinding" or plugging". The latter may be caused by several factors, one of the more common being crystal growth from the process itself. A good example of this can be found in processes where gypsum is encountered, e.g. on either horizontal belt or tipping pan filters in the production of phosphoric acid.

iii) Good cake discharge at the end of the filtration cycle:

Adequate cake release is a fundamental pre-requisite in efficient pressing operations, in maintaining a low down-time in the overall 'batch' time. The ability of a medium to discharge its cake depends very much upon the smoothness of the surface upon which the cake is residing, and hence upon the amount of fibrous material extending from the surface into the cake. Filter cake which adheres to the fabric inevitably leads to a reduction in process efficiency, either by way of a reduction in effective filtration area or where the cakes require to be removed with manual assistance, a longer and hence more costly cycle time. This particular problem has been addressed in recent years by manufacturers with the provision of high pressure wash jets and in filter presses, brush cleaning facilities. An understanding of the failure of the release mechanism follows consideration of the balance between the forces causing adhesion of the cake to the medium and the discharge forces. The adhesion of particles dispersed in liquids is mainly the result of electrostatic and Vander Waals interactions; chemical bonding also plays in important role. The effectiveness of discharge will depend on:

1) The strength of the bond between the cake and cloth is influenced by cake stickiness and yarn/weave characteristics. The bonding force depends on the mode of deposition of the first layer.

2) The internal strength of the cake. If the cohesion of the latter is less than the adhesion to the cloth, the cake will fail internally and leave solids on the cloth. The moisture content will vary across the depth of the cake.

3) The applied discharge force (e.g. gravity discharge from a vertical surface).

iv) Low cake moisture content:

This is particularly important with respect to processes where the cake has to be thermally dried. The high cost of thermal energy makes it imperative to express the maximum amount of moisture from the slurry by mechanical means. Similarly, where filter cakes are to be transported for landfill, the moisture content again has to be controlled to a low level to meet local statutory requirements. Both fabric and equipment have a part to play in this context.

v) Filtrate Throughput:

Throughput is the amount of fluid able to pass through a filter prior to plugging. The maximum throughput in the minimum time and with minimum resistance is perhaps one of the process engineer's most important objectives, the operation frequently being crucial to the balance of the total production cycle. Once again equipment parameters play a leading role in this subject.

vi) Filtrate clarity:

Whilst acknowledging that in most cases the role of the filter fabric is to achieve maximum separation of solids from liquids, absolute filtrate clarity may or may not be critical to the operation. The next destination of the filtrate and/or the ability of the process engineer to recirculate until satisfactory clarity is obtained need to be established and balanced against throughput requirements. Similarly in certain screening operations the fabric is designed to capture particles only of a specific size.

c) Equipment Considerations

In respect of equipment considerations the ideal filter fabric will, in simplistic terms, provide a long trouble-free performance. Equipment considerations again focus on the cleaning mechanisms and in particular, the forces applied by them. The following factors are considered while selecting the equipment for filtration process:

i) Cloth Shrinkage

Shrinkage of the cloth can produce severe problems particularly with plate and frames and the larger recessed plates. Repeated cloth washing/drying cycles aggravate the shrinkage problems, particularly for polyamides. It is sometimes recommended not to dry out cloths, after washing, storing wet if possible. Pre-shrinkage of fabrics is, therefore, widely practiced in order to retain dimensional stability in service.

Pre-shrinkage can be effected in a number of ways:

a) Use of hot (boiling) water with medium in a relaxed state

b) Heat setting in an oven, with fabric under tension in warp and weft direction. This will maintain the original levels of cloth porosity, permeability, etc...

c) Oven treatment of relaxed cloth or mild tension feeding through the oven

The shrinkage process has to be carefully controlled, in view of the large structural changes of up to 15% shrinkage which can ensue in relaxed conditions. Spun staple yarns are reported to shrink less than comparable fabric woven from filament yarns.

ii) Cloth Stretching

Absorption of liquids causes swelling of fibres and yam. The increase in fibre diameter and length causes dimensional change in the cloth, with serious consequences for closely fitted plates. This is particularly true for those yarns with poor absorption characteristics (nylon). The latter may absorb up to 4% by weight; this may be compared with 0.4% for Terylene.

Automation, which may involve discharge of a relatively thin cake by peeling apart the cloth and solids, usually also involves high-pressure squeezing during the dewatering cycle. The complete cycle of filling, pressing, air blowing, discharge and cloth washing may be short (5-6 minutes) in the modern auto-variable chamber unit. Thus the cloth may be subjected to as much stress in days, in modern units, compared with months in manually operated systems.

All these considerations call for cloths of great strength whilst retaining high-level filtration characteristics. For example, a traditional strong polyester fabric suitable for use on large plates (40x40 in) in slow or manual systems would have, say, breaking load figures of 900 N/cm (warp) and 350 N/cm (weft). These figures may be compared with cloths of the same size in mechanized units: 1800 N/cm (warp) and 1600 N/cm (weft), respectively.

iii) Filter Size

The size of a filter needs to be selected with regard to the acceptable pressure drop and the cycle time required between successive cleaning and element replacement. This is closely bound up with the type of element and filter medium employed. In conditions of heavy contamination, a filter element with high retention properties may clog too quickly for economic use, calling for a much larger size than normal, or alternatively a different type of element with better collecting properties, so that clogging is slowed down.

d) Cost

Although considerable technology goes into the production of a filter cloth - both in production and fabrication - and although it may be of vital importance to the success of the operation or the quality of a product, it is often acknowledged that the cost contribution of the filter fabric to the final product cost is extremely low. Filter fabrics are expected to provide the longest possible life before failure due to blinding or mechanical/chemical damage.

Yarn Construction and Properties:

a) Monofilament:

Monofilaments are single filaments extruded from molten polymer through a spinneret and then subject it to drawing to orientate the molecules and thus provide the thread with the desired stress strain characteristics. The monofilaments are usually round in cross-section although other profiles are possible.

Table 5.3 Typical Filtration Characteristics of different fibre forms

	Maximum Retention	Maximum production	Maximum Cake Moisture reduction	Maximum Cake Discharge	Maximum Life	Maximum Resistance to Blinding
Fibre Form	Spun staple	Monofilament	Monofilament	Monofilament	Spun staple	Monofilament
	Multifilament	Multifilament	Multifilament	Multifilament	Multifilament	Multifilament
	Monofilament	Spun staple	Spun staple	Spun staple	Monofilament	Spun staple

Fabrics produced from monofilaments are characterised by their resistance to blinding, their high throughput and their ability to discharge filter cakes cleanly and efficiently at the end of the filtration cycle.

b) Multifilament

Multifilament are extruded and orientated in much the same way as monofilaments although on this occasion the spinneret, contains a large number of much smaller apertures. The diameter of the individual filaments in this case is usually of the order 0.03mm. Extrusion is followed by twisting to bind the filaments together. The twist makes the filament assembly slightly stronger, more rigid and, if a high twist level is used, can alleviate the tendency of the yarn and hence the fabric to blind. Multifilament fabrics possess greater collection efficiency, higher strength and greater flexibility than monofilament fabrics are nevertheless more prone to blinding than the latter. Again, the relatively poor mechanical properties of monofilaments such as stretching wrinkling or tearing, promoted an interest in high-twist multifilament cloths woven in such a way as to ameliorate the associated problems of cake release.

c) Staple yarns

Staple spun yarns are produced from short fibres using spinning technologies which were developed for the processing of natural fibres such as cotton and wool. After extrusion the fibre length is cut to 40-100mm depending on which the staple spinning system is employed. Woollen spun yarns provide greater throughput, are more efficient and less prone to blinding than either multifilament yarns or yarns processed on the cotton spinning system. On the other hand, as with multifilament yarns, the blinding resistance of staple spun yarn is significantly inferior to that of monofilaments. High-twist yarns are advantageous if compressed air is used for cake discharge.

Twist	% Flow through yarns
1.5-3.0	95-98
15	70
35	2

Table.5.4 –Effect of Twist of Yarn on Particle Flow

d) Fibrillated Tape Yarns

These yarns are produced from narrow width polypropylene films which are converted into relatively coarse filaments by splitting the film either with special cutters or pins, hence the alternative term "split film yarns". These yarns find limited use in filtration, and are used mainly in the form of coarse, open weave structures providing support and drainage for the primary filter cloth.

Fabric Construction and properties
Woven Fabric Filters

Fabrics make up the largest component of filter media materials. They are made from fibres or filaments of natural or synthetic materials, and are characterized by being relatively soft or floppy, lacking the rigidity of dry paper, such that they would normally need some kind of support before they can be used as a filter medium.

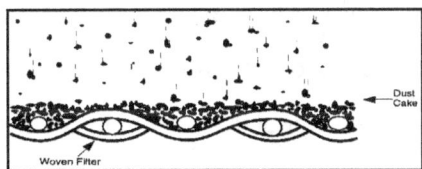

Fig.5.9 – Woven Fabric Filter

The structure of woven filter cloths depends on the type of yarn used in weaving. Yarns are available in several forms: monofilament, multifilament, staple fibre and mixtures of the same. The yarns in these cloths are composed of solid polymeric material (polypropylene, polyester, polyamide, etc.). Cloths of different patterns are produced on the loom by varying the manner in which the warp and weft yarns are woven together. The plain-weave monofilament cloth has been produced by warp and weft yarns of the same diameter, woven together in a simple one-under, one-over pattern.

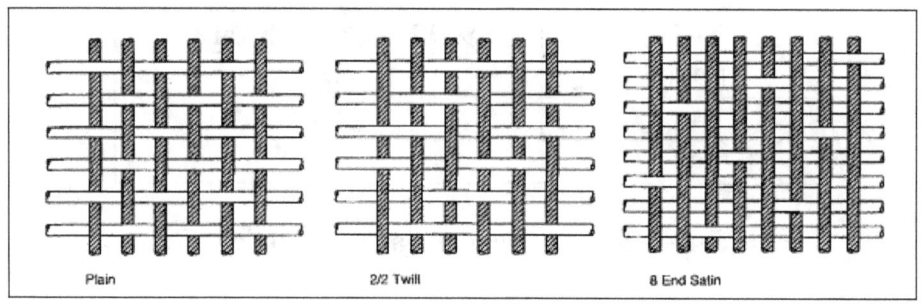

Fig.5.10- Types of Weaves in Woven Fabric Filter

These cloths are available in a wide range of pore sizes from 5000 to about 30 μm, the lower limit being determined by the size of fibre available for the weaving process. These cloths are characterized by pores of an open type which create little flow resistance and many applications are found in areas where high flows are required, e.g. in oil, paint, and water filtration and screening. Such cloths are readily cleaned by back-flushing. The surface of the fabric can be modified by finishing; is involves heat treatment and calendaring in order to flatten the surface, reduce pore size, but preserve the construction by way of reducing any tendency to shrink or stretch in service.

Unfortunately, these fabrics are light in weight and would be easily damaged if used directly in pressure filters. Thus the advantages of high throughput and ease of cleaning must be weighed against the fragile nature of the medium. Modern trends are to produce composite weaves from fine and coarse monofilaments in the production of a surface layer with good release properties and non-blinding characteristics. The underside layer of coarser fibres provides support, assist drainage and promote attachment or caulking of the cloth onto the filter platform. In order to produce apertures finer than mentioned above, changes in the weave are available to alter the size (and shape) of the cloth pores. Cloth of this type, e.g. sateen weave possesses very smooth surfaces which are optimal for cake release.

Table.5.5 Effect of weave type on properties of filter fabric

Property	Weave Type		
	Plain	Twill	Sateen
Rigidity	Best	Satisfactory	Poor
Bulk	Poor	Best	Best
Initial flow rate	Poor	Good	Best
Retention Efficiency	best	Satisfactory	Poor
Cake Release	Good	Poor	Best
Resistance to blinding	Poor	Good	Best

Table 5.6 Effect of Fabric weave on Filtration characteristics

	Maximum Retention	Maximum production	Maximum Cake Moisture reduction	Maximum Cake Discharge	Maximum Life	Maximum Resistance to Blinding
Weave pattern	Plain	Satin	Satin	Satin	Twill	Satin
	Twill	Twill	Twill	Twill	Plain	Twill
	Satin	Plain	Plain	Plain	Satin	Plain

Thus the more open weave fabrics will be superior in nonblinding characteristics, but may have poor particle retention. The latter will improve in the order monofilament < multifilament < staple fibre.

The relative amount of flow through and around the yams in such cloths will depend on the degree of twist imparted to the yam and the size of the apertures between yams. The aperture size will, in turn, depend on the weave pattern; plain, twill, sateen, etc... Swelling of fabrics can change the nature of flow in that closure of the cloth pores can force an increasing amount of flow through the yarns.

Properties of Woven Filter Fabric:

Three properties by which a filter fabric medium may be judged are:

- The permeability (or, inversely the resistance) of the clean, unused medium
- The particle-stopping power of the medium
- The permeability (or resistance) of the used, or deposited medium

The filtration process involves two principal resistances: (a) the resistance of the filter cake "α" and (b) the resistance of the medium "R_m". At high levels of α, e.g. greater than 1×10^{12} m/kg (characteristic of sludge-like material) changes in R_m, have little influence on overall productivity-at least in the range $1 \times 10^8 < R_m < 1 \times 10^1$. Thus a partially blinded medium may still function quite satisfactorily in a system controlled by α.

a) Filter permeability:

The permeability is the reciprocal of the resistance to flow offered by the filter thus, high permeability represents a low resistance and vice versa. Permeability is usually expressed in terms of a permeability coefficient, which is directly proportional to the product of flow rate, fluid viscosity and filter medium thickness, and inversely proportional to the product of filter area and fluid density, which gives the permeability coefficient the dimension of a length.

For a given flow rate, an increase in filter area will reduce the pressure drop across the filter, because the amount of fluid flowing per unit of filtration area is decreased (pressure drop is inversely proportional to filter area).

The operating temperature of the fluid will affect the pressure drop across the filter because the fluid viscosity will change. A less viscous fluid will experience less resistance to flow through the medium, and so a lower pressure drop will be needed to drive it. As a result, pressure drop is inversely proportional to temperature, with a decrease in temperature causing a rise in pressure drop.

The effect of prolonged filtration time is to produce a cumulative build-up of collected solids on or in the filter medium, thus reducing permeability (and increasing flow resistance) in direct proportion to the amount of solid collected.

Table 5.7 Typical values of air permeability of fabrics:

Fabric Type	Air permeability in m^3/m^3sec
Nylon Multifilament	0.030-1.52
Polypropylene monofilament	0.015-1.52
Polypropylene multifilament	0.005-0.508
Nonwoven cloth	0.002-1.27

Thus a high permeability is taken as an indication of high porosity and, in turn, low particle retentivity.

i) Multifilament Cloth Permeability:

In multifilament cloths, fluid flow may occur through or around the permeable yarns. If we define B_0, as the permeability of the porous yarns, and B_1 as the permeability of the cloth if the yarn were solid, i.e. monofilament, it may be shown that:

$$\Omega = \frac{B}{B_1} = \left(1 + 1.34 \left(\frac{B}{B_o} \right)^2 \text{ for } \frac{B_o}{d_y^2} < 0.0017 \right)$$

where B is the overall permeability of the cloth and d_y is the yarn diameter. The Ω index has been shown to vary in the range $1 < \Omega < 20$ within the order of accuracy of the experimental measurements necessary for the determination of B and B_0.

$$\Omega = \frac{\text{Permeability of cloth}}{\text{Permeability of cloth composed of monofilament yarn}}$$

ii) Monofilament Cloth Permeability

In the monofilament area much more success in correlating permeability with cloth structure has followed the suggestions of Pedersen [1969], who adopted orifice-type formulae to correlate pressure-drop-flow information for various weave patterns.

A discharge coefficient was defined as:

$$C_D = \left(\frac{v^2}{2\Delta P} \frac{(1 - a^2)}{a^2} \right)^{0.5}$$

where a, the effective fraction open area of the pore is: $a = A_0\,(ec)\,(pc)$ in which,

(ec) = warp yarns per centimetre

(pc) = weft yams per centimetre

A_0 = effective area of orifice

V = flow rate

ΔP = Pressure drop

The discharge coefficient was anticipated to be a function of the Reynolds number within the fabric.

Nonwoven Filtration Media:

Today, the filtration industry worldwide is growing at 2 to 6 percent per year above gross domestic product. Nonwoven fabrics have seen tremendous growth with penetration into a number of filtration industry end-use market segments. Nonwovens offered a cost effective alternative and often a distinct technical advantage by the basic attributes of the nonwoven construction. The factors such as fibre diameter, orientation, packing density and web weight will determine the filter media properties. Nonwoven fabric and membrane filtration media together dominate the filtration media market, with more than 90-percent combined market share in terms of roll goods filtration media volume in comparison to all other material forms. Typically, nonwoven fabrics add backup support and/or mechanical strength to comparatively weak membrane media, allowing membranes to function at peak performance.

Nonwoven fabric filtration media have dominated in applications such as coolant filtration, bag house filtration media, vacuum cleaner bags and many heating, ventilating and air conditioning (HVAC) applications. In these and other applications, nonwovens are highly price-competitive. Air applications consume approximately 65 to 70 percent of the nonwoven filtration media, with liquid uses consuming the remaining 30 to 35 percent. Overall, 75 percent of synthetic nonwoven media go into commercial markets, such as manufacturing facilities, offices, theaters, hospitals, cruise ships, casinos and other such markets; with about 25 percent found in residential and general consumer air filters.

The Nonwoven fabric media vary by materials of construction, processing method and performance characteristics. They can be classified into two distinct types based on their method of formation. The first method is dry laid processes, which includes carded, needled, spunbond and melt blown media. The second process uses a wet laid formation, which is generally done on a paper machine. Each process produces a media with unique properties that have advantages in different applications.

Dry Laid Media:

Dry laid process is a technology in which fibres are uniformly dispersed in air, deposited and transported onto a continuous moving fine mesh screen and then made to form a mat as a result of filtration. Dry laid processes generally produce media with nominal ratings that are low cost and have high dirt holding capacities. Melt Blown media are one of the most versatile nonwovens for liquid filtration. Melt blown media is generally composed of a continuous network of self-bonded polypropylene, polyester or nylon microfibers produced with a controlled fiber uniformity and density. The resulting media has a uniform porosity, does not shed fibers and contains no binders, adhesives or surfactants. Melt blown media have nominal ratings from 1μm to 50μm. Nonwoven melt blown and spunbond fabric along with nonwoven glass filtration media are the principal air filtration media for HVAC.

Wet Laid Media:

Wet laid Nonwovens technology involves uniformly dispersing short cut fibres in water, transporting the slurry onto a continuous moving fine mesh screen called the wire, and then forming a mat as a result of the removal of water. The web then undergoes further water elimination through drying. A major objective of wet laid nonwoven manufacturing in filtration is to produce structures with known pore size and filtration characteristics.

High-efficiency particulate air (HEPA) wetlaid glass nonwoven filtration media represent 90 million to 100 million square meters nonwoven market. Air filters are found in end-use markets from general dust filtration to high-efficiency filtration in many different configurations. These filters are rated by a Minimum Efficiency Reporting Value (MERV) standard, which rates filters from 1 to 20 in terms of their degree of efficiency. At the high end, MERV 17- to 20-rated HEPA filters are typically used in situations that require absolute cleanliness for the manufacture of microchips, liquid crystal display screens, pharmaceutical production and microsurgery in hospital operating rooms. HEPA filters are primarily constructed from wetlaid glass nonwoven filtration media. Wetlaid cellulosic and spunbond polyester media that range from 200 to 300 grams per square meter (g/m ^{2}) are used in pleated dust collection cartridges. Pleated cellulose- or polyester-based filters offer significantly greater surface area than needle felt filter bags for a given space as an alternative filter configuration in bag house applications.

Characteristics of Nonwoven Filter media:

The high porosity of nonwovens is an advantage in producing a high retention capacity for depth filters. These filters usually contain bonded fibres. Thus the wet strength and overall resistance to fibre shedding, etc., can be improved by sealing the fibres, one to the other, to produce a rigid network. Bonding can be effected by the inclusion of adhesives or by heat setting. The latter is inherent in spun-bonded fabrics. These involve the extrusion of molten polymer into cylindrical filaments which are dispersed by hot gas flow into a tortuous, random array. The fibre mixture may include a small proportion of low melting point material. A wide variation in fibre diameter exists.

Examples are:

i) 10 μm Polyester (1.3 dtex where the latter unit is the weight in grams at 10 000 m of fibre).

ii) 40 μm Polypropylene (13.3 dtex).

iii) 30 μm Cellulose (may be fibrillated to produce fine fibre attachments or fibrils).

iv) 0.03-8 μm Glass (100% glass media used in laboratory liquid separation and in gas filtration).

Both the permeability and filtration characteristics of nonwovens are dependent on the felt porosity and fibre diameter. A medium which has been heavily calendared on both sides will possess the lowest porosity. Surface treatments and/or use of laminations of different porosities are aimed at improving cake filtration performance and cake release. Generally speaking, the filtration efficiency at a particular particle size is inversely proportional to the fibre diameter, other factors being the same.

Factors Influencing the Air Permeability of Non Woven Filter:

a) Effect of Fabric Weight

Fabric weight which is generally measured in gm/m^2 play an important role on performance of non woven filter fabric. The air permeability decreased with the increase in fabric weight due to more number of fibers per unit area. Also due to increased fabric density the resistance to air flow increases. With increase in fabric weight the pressure drop increases thus improving the filtration efficiency. The tenacity at break increased with the fabric weight in both bias and cross direction. This is due to the increase in no. of fibres in the web, leading to increase in no. of vertical loops and density and entanglement, thus causing less freedom of fibre movement and greater frictional resistance. But the breaking elongation decreases gradually in both bias and cross direction with the increase in fabric weight. The abrasion resistance of the fabric increases with the increase in fabric weight due to the increase in compactness and density of fabric. The bursting strength also increases with the increase in fabric weight due to the increase number of fibres which play an important role in resisting the bursting pressure.

b) Effect of Fabric Density and Thickness

The air permeability decreases non-linearly as thickness or fabric density increases. The density had a more significant influence on air permeability than either thickness or fibre size. There can be no general correlation between porosity and permeability because the permeability of a material is influenced by the capillary pressure curves. Air resistance increased with fabric thickness and fabric weight per unit area, but decrease with fibre fineness.

c) Effect of Synthetic and Cellulosic Fibers Blend:

This figure shows the particle retention of the different media types using AC test dust. There are no clear trends with the fine particle retention. The cellulose/synthetic blends and the 100% synthetic media have similar particle retention and the synthetic composite media is slightly lower but the cellulose media had poor particle retention. The coarse particle retention does show a trend of improving as the synthetic content in the media increased. The dirt holding capacity performed as expected based on the standard paper airflow or resistance testing. The dirt holding capacity improves with the addition of 15% synthetic fiber to a cellulose based media.

Finishing Treatments

a) Heat Setting:

Heat setting is a dry process used to stabilize and impart textural properties to filter fabrics. When fabric filters are heat set, the cloth maintains its shape and size in subsequent finishing operations and is - stabilized in the form in which it is held during heat setting (e.g., smooth, creased, uneven).

b) Singeing:

Fabrics produced from short staple fibres naturally posses a fibrous surface, which, in some cases, can impede cake discharge through mechanical adhesion of fibre to cake. Singeing is predominantly carried out on textile fabrics to achieve a smooth and fibre-free surface. Singeing enhances the smooth cake release.

c) Calendaring:

Calendaring is a process where fabric is compressed by passing it between two or more rolls under controlled conditions of time, temperature and pressure in order to alter its handle, surface texture and appearance. It is done to improve the fabric's surface smoothness (for better cake discharge) and also to regulate its permeability and hence improve its collection efficiency.

d) Raising or Napping

Raising process is designed actually to create a fibrous surface, normally on the outlet side of the filter sleeve, to enhance the fabric's dust collection capability. Raised fabrics may comprise 100% staple-fibre yarns or a combination of multifilament and staple-fibre yarns, the latter being woven in satin style in which the face side is predominantly multifilament and the reverse side predominantly staple.

e) Antistatic Finish

Electrostatic charges within a filter dust cake can develop by friction during the processing and movement of gases and fine dust particles. These static charges can be sufficient to generate sparking giving an ignition source to initiate an explosion of combustible gases and dust particles. Anti-static filter media have electrically conductive substrates designed to safely dissipate these charges. The unique microporous structure of the PTFE membrane comprises millions of randomly connected fibrils giving an effective pore size many times smaller than can be seen by the naked eye. The result is a surface able to capture very fine particulate, while allowing air and static charges to freely permeate the media. Anti-static media is used to dissipate electrostatic charges where explosive hazards exist or where charged dust particles resist release from non-conductive filter media.

Nanofiltration

Nanofibers can be defined as fibers with a diameter of less than 1mm or 1000nm. The majority of nanofibers are produced by the electro spinning process—a process that has been used to spin fibers since the early 1930's. In this process a charged polymer melt is extruded through a small nozzle. The charged solution is drawn toward a grounded collecting plate. As the jet of charged melt travels, the solvent evaporates, leaving a non-woven nanofiber mat on a substratum. The nonwoven webs of fibers formed through this process typically have high specific surface areas, nano-scale pore sizes, high and controllable porosity and extreme flexibility with regard to the materials used and modification of the surface chemistry of the fibres. The process can be altered to produce fibers with different diameters. Nanofibers are characterized as having a high surface area to volume ratio and a small pore size in fabric form. The high surface area to volume ratio and small pore size allows viruses and spore-forming bacterium such as Anthrax to be trapped. Filtration devices and wound dressings are just some of the applications in which nanofibers could be utilized.

Studies have shown that a low fibre diameter allows a filter element with similar operational characteristics but with much higher filtration performance. Any process that is dependent on surface area, such as active filtration will benefit from the incorporation of nanofibres into the process. The first successful commercial use of electro spun fibre was claimed by Donaldson Inc., for filtration elements. Donaldson Co. Inc., have been developing electro spun nanofibre filtration elements for dust collection, gas turbine air filtration and air filters for heavy-duty engines. Active filtration implies that the entrapment method is based on chemical attraction rather than simple physical entanglement. The advantages of this method are a lower resistance to flow across the filter element, and the possibility of selectivity so that particular elements can be removed during filtration.

Pleated dust collection and engine air-intake filters have a lightweight cover of synthetic nanofibers over a base substrate of a wetlaid cellulosic or polyester nonwoven in a growing number of applications. The nanofibers are as fine as 200 to 300 nanometers (nm) in diameter, with the amount of nanofiber add-on being quite thin in cross-section and typically weighing less than 1 g/m^2 to 2 g/m^2. The nanofibers are laid down over what will become the upstream side of the substrate using an electro spinning process, and in one case, an ultrafine melt blown process. These nanofibers create a labyrinth of fibers with pores finer than particles in the incoming air stream. Particulate deposits and resides on the surface of the fine nanofiber web, allowing the user to clean the filter by shaking off loose particles from the surface or by using an automated clean-air back-pulse system.

References:

1. A. Rushton, A.S. Ward, R.G. Holdich, Solid-Liquid Filtration and separation technology, VCH Publication, 1996.

2. Derek B. Purchas, K. Sutherland, Handbook of Filter Media, Elsevier, November 2002.

3. Ken Sutherland, Filters and filtration hand book, 5th edition, B-H publications, 2008.

4. Irwin Marshall Hutten, Handbook of nonwoven filter media. 2007.

5. Edward C. Gregor, "Nonwoven fabric filtration", The Textile World, March/April 2009.

6. U J Patil, Prafull P Kolte, Filtration in Textile: A review, Indian Textile Journal May 2011.

7. Christoper Shields, Submicron Filtration Media, Nonwoven perspective, INJ Fall 2005.

8. R J Wakeman, Filtration Dictionary and Glossary, The Filtration Society (UK), 1985.

9. H W Ballew, Basics of Filtration and Separation, Nuclepore Corporation, 1978.

10. D B Purchas 'Art, science and filter media', Filtration O Separation, 17(4), 3 72-6, 1980

11. F M Tiller Theory and Practice of Solid-Liquid Separation, University of Houston, TX, 1978.

12. A Rushton and P V R Griffiths, 'Filter media', Filtration Principles and Practices, Part 1 (ed. Clyde Orr), Chapter 3, Marcel Dekker, 1977.

13. A Rushton, 'Effect of filter cloth structure on flow resistance, blinding and plant performance', The Chemical Engineer, No. 237, 88, 1970.

14. T C Dickenson, Filters and Filtration Handbook, 4th Edition, Elsevier Science, 1997.

15. S Ehlers 'The selection of filter fabrics re-examined', Ind. Eng. Chem, (International) 53(7), 552-6, 1961.

16. E Hardman, 'Some aspects of the design of filter fabrics for use in solid/liquid separation processes', Filtration & Separation 31 (60), 813-18, 1994.

17. R Krcma, Manual of Nonwovens, Textile Trade Press, Manchester, UK, 1971.

18. E Mayer and H S Lim 'New nonwoven microfiltration membrane material', Fluid-Particle Separation journal, 2(1), 17-21, 1989.

19. H N Sandstedt, 'Nonwovens in filtration applications', Filtration & Separation, 17(4), 358-61, 1980.

www.ingramcontent.com/pod-product-compliance
Lightning Source LLC
Chambersburg PA
CBHW081243170526
45165CB00009B/3168